Inside the World of Drones

DRONES AND WARFARE

Jeff Mapua

ROSEN
PUBLISHING

New York

Published in 2017 by The Rosen Publishing Group
29 East 21st Street, New York, NY 10010

Copyright 2017 by The Rosen Publishing Group, Inc.

First Edition

All rights reserved. No part of this book may be reproduced in any form without permission in writing from the publisher, except by a reviewer.

Library of Congress Cataloging-in-Publication Data

Names: Mapua, Jeff, author.
Title: Drones and warfare / Jeff Mapua.
Description: New York, NY : Rosen Publishing, [2017] | Series: Inside the world of drones | Includes bibliographical references and index.
Identifiers: LCCN 2016029556 | ISBN 9781508173373 (library bound)
Subjects: LCSH: Drone aircraft. | Air warfare—History—21st century.
Classification: LCC UG1242.D7 M335 2016 | DDC 358.4/14—dc23
LC record available at https://lccn.loc.gov/2016029556

Manufactured in China

CONTENTS

INTRODUCTION 4

Chapter 1
RISE OF THE MACHINES 7

Chapter 2
DRONES IN ACTION 18

Chapter 3
THE ETHICS OF DRONE WARFARE 30

Chapter 4
DRONE WARFARE OF THE FUTURE 43

GLOSSARY 54
FOR MORE INFORMATION 55
FOR FURTHER READING 58
BIBLIOGRAPHY 60
INDEX 62

INTRODUCTION

A United States Air Force pilot spots his targets walking into a building in a village in Afghanistan known to harbor antigovernment insurgents. The pilot is calm and collected as he controls his secret weapon. It took a year of training to learn to wield this new weapon in the U.S. military's arsenal: an unmanned aerial vehicle (UAV), commonly referred to as a drone.

The drone closes in on the target, but the pilot is not in the line of fire. He's not even in the same country. While the drone nears its destination, the airman pilots the vehicle from the other side of the world, safely stationed in a military base somewhere in the United States.

The pilot is seated in what could be mistaken for a fully-furnished video game setup. The "cockpit" is a desk with an array of computer screens showing the pilot a series of maps, video feeds, and gauges. The pilot uses a joystick to maneuver the drone, while another set of controls governs the throttle.

Meanwhile, the men in Afghanistan keep their voices low, trying to listen for an incoming drone. They've heard the stories from drone strike survivors, know what power drones carry, and that they can strike as suddenly as lightning, causing catastrophic damage. They could be under surveillance at that very moment.

In the United States, the order to fire is given, perhaps even coming directly from the president. With a push of a button, a laser-guided missile races through the air headed

Unmanned aerial vehicles like this MQ-9 Reaper are capable of carrying out military missions while keeping soldiers and other military personnel out of harm's way.

straight for the men in the building. The loud hissing sound the missile creates alerts the men that they are in immediate danger, but it is already too late. The building is destroyed, debris flying chaotically onto the streets below.

Unmanned aerial vehicles, or drones, have evolved into the United States' preferred weapons. They are favored because they keep live soldiers out of harm's way, reducing the number of American casualties. Drones can carry out surveillance missions that may be too physically taxing, or too time-consuming for humans to perform. As technology advances, drones will even shrink down to sizes that open up a whole new micro level of espionage. UAVs invisible to the naked eye are on their way, and may perhaps be deployed sooner than we think. It may seem that drones will be ubiquitous, or everywhere, in the future.

However, there are many objections to drone use in warfare. The use of drones has raised many legal questions in the international community, with many claiming that their use breaks international laws—at least the way they are used specifically in the so-called War on Terror. Not every drone mission goes as planned. Opponents accuse the U.S. military and intelligence services of indiscriminately killing too many innocent civilians. They point out that these deaths lead to more anger toward countries like the United States and create more armed backlash against Western nations and their allies.

Drone proponents counter that even more civilians would have died had they not carried out certain attacks. The topic is the subject of myriad books, articles, news specials, and movies. Still, with no end to drone programs in sight, a huge part of the future of warfare looks to be unmanned and airborne.

CHAPTER 1

RISE OF THE MACHINES

Humans have always relied on technology to procure food and to make war. Around 400,000 BCE, humans used the spear to hunt for food. The earliest arrowheads found date back to around 20,000 BCE, and the projectile weapon became more widespread over time. Weapons technology evolved along with human society. In 1415 CE, the use of the longbow decided the outcome of the famous Battle of Crecy, in which English archers beat a French army that outnumbered their forces five to one.

Eventually, the bow and arrow gave way to guns, invented during China's Ming dynasty in the 13th Century CE. In 1914, the British army introduced tanks and turned the tide in their favor during World War I. The unmanned aerial vehicle, or drone, is the latest in the long line of war-making advances. It has ushered in a whole new era of warfare.

A HISTORY OF UNMANNED AERIAL VEHICLES

Looking back at the military origins of drones, they arose, surprisingly, during the Civil War, when aerial military surveillance included the use of hot-air balloons. The manned, gas-filled balloons gave soldiers an ability to see beyond what they could on the ground. Both sides of the Civil War used them for espionage, tracking troop movements, and helping direct artillery fire.

Kites were even used during the Spanish-American War in 1898. Reconnaissance photos were taken by fastening a camera to the kite and flying it over a targeted area. Kites and hot-air balloons eventually gave way to airplanes in World War I, when they were used both for surveillance and as weapons. In World War II, the early drone materialized in the form of radio-controlled missiles designed by Nazi scientists and used on targets in England.

Radio-controlled B-24s were also used by the Allies to bomb Germany. In the space race during the Cold War (after 1945) between the United States and the Soviet Union, scientists invented remote-controlled objects in the form of satellites and remote-controlled rockets and missiles.

Until later, drone warfare seemed like a plot device from science fiction. But during the Vietnam War, myth became reality when unmanned aerial vehicles were deployed. The first drones were used for intelligence, surveillance, and reconnaissance. During the Vietnam War, drones flew around 3,500

The Global Hawk shown here made history in 2001 when it became the first UAV to fly 7,500 miles without refueling, during a trip it made across the Pacific Ocean.

missions. The drones were able to capture still images of the battlefields. Attack drones were developed during this time, but the technology was not reliable and accurate enough to allow users to identify and neutralize targets effectively.

Israel developed their own drone technology, with its military using them for various missions, including for surveillance and as decoys. Unlike drone use in the Vietnam War, attack drones were used by Israel in targeted assassination missions.

The United States then purchased several of the Israeli drones to use for their own missions including those in the First Gulf War. The new drone model, called the Pioneer, was used in more than three hundred missions in the Persian Gulf. In one instance, Iraqi soldiers surrendered to a Pioneer by waving white flags to the circling drone overhead.

The United States used drones in various missions through the 1990s, and by 2000, the United States Department of Defense, also known as the Pentagon, wanted to expand the drone program. Their goal was to make one-third of all U.S. aircraft unmanned by 2010.

A drone model called the Predator successfully located Osama bin Laden in Afghanistan in 2000. Bin Laden and Al Qaeda were tied to terrorist bombings in 1993 and 1998. However, carrying out a raid on their locations proved to be too complex and risked too many lives of troops and civilians. Security officials wanted to turn drones into weapons. In 2001, the 9/11 terrorist attacks against the United States gave the government the incentive to do just that. The War on Terror prompted the military to make drones into weapons rather than strictly intelligence-gathering tools.

Over time, drones were able to link into the global telecommunications system. No longer were drone operators required to be on the ground near the drones. Pilots could now control them from anywhere in the world. Today, drone pilots are usually completely shielded from physical danger when operating them in conflict zones.

HOW THEY WORK

Drone technology has evolved over time and become more sophisticated. Drones resemble airplanes, albeit smaller, and without windows or room for a pilot and crew. Instead, the nose of the drone is used to store a number of various sensors and its navigation systems. These can include color or black-and-white video cameras, image intensifiers, radar, infrared imaging for low-light conditions, and lasers for targeting. These sensors and cameras send data and visuals to the personnel in charge of the mission. Attack drones are armed with laser-guided missiles.

Drone pilots are located in ground-control stations. These are typically in the same areas or regions as the drones themselves. Pilots here control landing and takeoff. Once airborne and out of the ground-control station's line of sight, the drones can be flown by pilots located at a base anywhere around the world. The drones communicate with satellites to receive instructions from pilots and update their locations via the Global Positioning System (GPS). The cameras are also able to send images or video to the troops on the ground.

UNDERGROUND AND UNDERWATER

Drones are not just for the skies. Various industries have found uses for drones underground for use in projects that involve subway systems or mining. Drones are used to capture images and video of tunnels and explore areas too dangerous for humans to investigate. Scientists designed a flying robot to fly through a German coal mine, flying through

A drone pilot flies a practice mission at Creech Air Force Base in Indian Springs, Nevada. The same facilities are used to pilot drones overseas from the safety of the western United States.

swirling dust, dark tunnels, and around various obstructions. It was able to fly on its own without human guidance. The hope is for these drones to assist in search and rescue missions after a disaster.

The military will introduce an even more ambitious type of drone: the Anti-Submarine Warfare Continuous Trail Unmanned Vessel, or ACTUV. This massive drone is 132 feet (40 meters) long and weighs 140 tons (127 metric tonnes), about half the size of a blue whale. The main uses for the underwater drone will be for reconnaissance, resupply, and tunnel warfare. The Navy hopes that the drone will reduce their expenses, cutting per day costs from $200,000 to $20,000.

The ACTUV is designed to use sonar to detect its surroundings, which would include other submarines and enemy ships. Its designers expect its range to be thousands of miles with sixty to ninety days of operation without a crew member inside. If it passes tests, the ACTUV will see action beginning in 2017. The ocean provides even more uses for drones as unmanned boats are also in development.

If a drone loses contact with the satellites, it is programmed to fly in circles until the link can be reconnected. This is a major issue and can lead to catastrophe. If they are unable to reconnect before running out of fuel, they eventually crash.

TYPES OF DRONES

Not all drones are equal. Some are designed for warfare while others are equipped with surveillance equipment. There are those armed for both. Different branches of the U.S. military use different drone models to suit their needs.

SURVEILLANCE

The Air Force and the Army both use a drone called the RQ-4 Global Hawk. This drone is designed for high-alti-

Models like the MQ-1 Predator—like this one shown in 2008 at Kandahar Airfield in Afghanistan—are equipped with surveillance equipment like cameras and sensors.

tude and long-distance endurance reconnaissance and surveillance. It can fly as high as 60,000 feet (18,288 m) with a range of almost 10,000 miles (16,093 km). It is about the size of a Boeing 757 and can operate in any type of weather, day or night. Work on the Global Hawk began in 1995 as a technology demonstration. It has flown in support of overseas operations since November 2001. In the Department of Defense's terminology, the *R* in its name stands for "reconnaissance" while the *Q* designates the drone as unmanned.

The MQ-1 Predator, originally called the RQ-1 Predator, is an Air Force drone first launched in 1994, equipped with powerful cameras and sensors. It can go as high as 25,000 feet (7,620 m) and can stay in the air for twenty hours at a time. It is relied on for intelligence gathering, surveillance, identifying targets, and reconnaissance. It would later be weaponized with laser-guided Hellfire II missiles in 2002. These missiles have a range of up to 5 miles (8 km).

SEEK AND DESTROY

The Department of Defense directed the Air Force to support initiatives of operations overseas, and the Air Force proposed the MQ-9 Reaper. This drone is the larger, more powerful, successor to the Predator model. The *M* in its name designates the drone as "multi-role," meaning it is used in both surveillance and attack missions. The Reaper is able to destroy or disable time-sensitive targets with

Drones and Warfare

Naval officers show off the helicopter-like MQ-8 Fire Scout drone on the deck of the aircraft carrier U.S.S. Fort Worth in the Port of Busan, South Korea, in March 2015.

persistence and precision. It can elevate up to 50,000 feet (15,240 m), fly at speeds up to 230 miles per hour (370 kilometers per hour), and carry four Hellfire missiles. The Air Force plans to replace all Predators with Reapers by 2018.

The MQ-5 Hunter, developed in the 1990s, is an Army drone powered by two engines. It can fly at elevations up to 20,000 feet (6,096 m) and carries Viper Strike munitions. The MQ-1C Gray Eagle, another drone used by the Army, is its upgraded version of the Predator system. It succeeds the MQ-1 Warrior and can carry four Hellfire missiles. Designed for long missions, it can sustain flight for twenty-five hours.

OTHER FUNCTIONS

The Navy uses its own drones for its missions. The MQ-8 Fire Scout is different from the other branches' drones in that it looks like a helicopter rather than a plane. It is deployed in support of Special Operations forces, providing reconnaissance, situational awareness, and precision targeting support. The Navy will be able to track threats with the Fire Scout, and other models can carry payloads of up to 700 pounds (317.5 kilograms). It can take off and land vertically from ships at sea, climb to 20,000 feet (6,096 m), and fly 125 miles (201 km) for flights lasting around eight hours.

The QF-4 Phantom began life as a fighter jet. The Air Force went back and retrofitted old F-4 fighter jets into drones. Rather than using the drone for surveillance or attack missions, the Phantom is instead used for target practice and testing various weapons systems.

CHAPTER 2

DRONES IN ACTION

Determining the first use of a drone by the U.S. government largely depends on one's definition of a drone. If defined as unmanned aerial vehicles, then there have been blueprints for drones dating back to the Civil War and the era of hot-air balloons. Drones—defined as unmanned vehicles that fly rather than float—were developed during World War I, but the technology rarely worked or was rudimentary.

World War II saw the United States fly an early prototype of a drone. However, radio-control technology was relatively primitive. A pilot would physically fly the plane at takeoff, reach cruising altitude, then parachute out of the aircraft when controls were taken over by a pilot located elsewhere. The program was a disaster. President

John F. Kennedy's older brother Joseph was one of the program's first pilots and his plane exploded before he could escape. The type of drone used today would come much later.

Although unmanned vehicles are commonplace in warfare today, they were not always an available tool for the military. At some point, the U.S. Department of Defense had to develop, build, and test drone use in live missions. Once drones were approved for active use, they have been utilized in many different ways to accomplish a wide range of goals.

THE UNITED STATES' FIRST STRIKE

The modern drone, an unmanned vehicle with computer-assisted technology, began to appear as designs and prototypes in the 1970s and 1980s. The U.S. Air Force used drones called Fireflies in Southeast Asia. Their reconnaissance missions took advantage of their small size and long range. The experimental drone program did not expand, as its models were deemed too expensive to build. In the early 1990s, several Pioneer drones were purchased by the Navy for use in the Persian Gulf War. The drones were fitted with sixteen-inch (forty-centimeter) guns and were a success in reconnaissance.

The US military used Predator drones for reconnaissance purposes during the 1995 conflict in the Balkans (the

20 // Drones and Warfare

Mullah Omar, the former leader of Afghanistan's Taliban, is shown in a wanted poster. He was among the first targets of US drone warfare.

former Yugoslavia). These early drones had issues, including the fact that they had little defensive capabilities and therefore were easily shot down. The wings would also freeze and cause the drones to crash, and there were no facilities for drone pilots.

Following the terrorist attacks on September 11, 2001, the United States military used a drone in the first-ever combat strike by a remotely piloted aircraft. On October 2, 2001, the target was Taliban Supreme Commander Mullah Mohammed Omar in Afghanistan. The Air Force utilized a CIA Predator drone fitted with Hellfire missiles to drive terrorist group al-Qaeda and the Islamic fundamentalist political group the Taliban out of the country. While the military attacked Taliban forces with almost four hundred aircraft and more than thirty ships, the CIA-controlled drone was secretly flown into Afghanistan. US intelligence

The drone pilot shown here controls a US Air Force MQ-1B Predator unmanned aerial vehicle from a secret base in the Persian Gulf.

identified Omar's home and had a chance to strike first in the oncoming war. However, instead of firing on the building Omar had entered, the Predator destroyed a vehicle outside. In the ensuing chaos, Omar was able to escape.

RECONNAISSANCE

In the Army, unmanned aircraft systems operators, or remote pilots, are seen as intelligence specialists with reconnaissance as their major objective. Aside from air reconnaissance, they plan, analyze, and then undertake surveillance, targeting, and acquisition missions. For their drone flights, the job requires the pilots to prepare maps, charts, and intelligence reports, analyze aerial photographs, and use a variety of computer systems. In doing so, they are prepared to gather as much intelligence as they can without losing control of their drone.

Pilots must take ten weeks of Basic Combat Training and over twenty-three weeks of Advanced Individual Training that takes place in the classroom, in the field, and on the job. Aside from missions, operators are expected to perform maintenance on related equipment such as power sources and communications-related vehicles.

Piloting a drone has often been compared to playing a video game. A joystick is used to control the aircraft, spot the enemy, and take reconnaissance photos. According to military commanders, pilots who locate targets of interest on which actions can be taken are commended.

While the missions can be stressful, studies show that drone operators can also easily suffer boredom. Operators performed better when there was a lot going on, but they would miss details during slow moments of a mission. Unlike driving, where too much distraction leads to accidents, long periods of not seeing anything results in poor drone performance.

There are also psychological effects to operating drones remotely that have been studied and those still ongoing that seem to indicate that, much like fighters in live combat, drone operators can and do experience post-traumatic stress disorder (PTSD). Despite the idea that drone warfare is like gaming, being exposed to violent killing that they themselves

A small drone that was shot down is handled here by a member of the Mahdi army in Iraq, a Shiite militia loyal to the cleric Muqtada al-Sadr (the man shown in the poster behind him).

initiate may be taking a toll on drone pilots that has yet to be fully understood. A few pilots told Salon.com and other media outlets about feeling guilt taking life from afar. Others felt overworked and underappreciated, declaring that combat soldiers and pilots sometimes saw them as "Nintendo warriors."

FIGHTING TERRORISM

The War on Terror, and related military campaigns sometimes lumped in with it, are different from U.S. wars in the past in that instead of conventional warfare, the military utilizes counterterrorism and counterinsurgency. The use of drones has escalated as American officials grow weary of costly wars that require enormous manpower and financing. Drones and small-scale raids have proven to be useful, if complicated, tools in eliminating identified enemies. Combined with lower costs and better technology, President Barack Obama's administration pushed for more drone usage. In doing so, certain sectors of the intelligence community are becoming militarized as drones made for surveillance and reconnaissance are more often armed for air strikes.

What began during President George W. Bush's administration, which ordered fewer than fifty drone strikes, vastly expanded during Obama's tenure with more than four hundred strikes ordered by 2013, with a steady rate of engagement since. They are used to assassinate key leaders among insurgent movements and terrorist groups fighting Western nations and their allies, in places like Pakistan and Yemen. Drones are able to zero in on targets in areas too dangerous or tough to attack with troops on the ground.

Aside from targeted killings, drones are used in cutting communications between hostile groups and new recruits. Terrorist groups such as al-Qaeda and enemies like the Taliban are known to avoid using electronic communications, which could be tracked by enemy intelligence methods. Without a means to communicate, their leaders find it harder to coordinate missions and attacks. Additionally, training groups of new recruits is difficult. Large gatherings and encampments are likely to be spotted and/or attacked by drones.

DRONES IN AFGHANISTAN AND PAKISTAN

Drone use has proliferated in Afghanistan and Pakistan, especially in remote regions too dangerous to attack by traditional methods. By December 2001, President Bush came out strongly in favor of the Predator drone. In a speech at The Citadel, the military college, as reported by CNN, Bush pointed to the drone's ability to "circle over enemy forces, gather intelligence, transmit information instantly back to commanders, then fire on targets with extreme accuracy."

MEN WALK BY A MURAL IN DECEMBER 2013 IN SANAA, THE CAPITAL OF YEMEN, WHICH DEPICTS A U.S. DRONE AND THE CAPTION "WHY DID YOU KILL MY FAMILY."

Drones are able to fly through war zones and unstable regions where attempting to capture militants has proven inefficient or untenable. Many such regions, such as the frontiers of Pakistan and Afghanistan, are too dangerous for troops to invade, or their governments will not allow large foreign forces on their sovereign territory. Additionally, local governments sometimes work with militants and make it more difficult for U.S. forces to distinguish friends from enemies.

Drone campaigns overseas have been met with great controversy, especially from human rights advocates, and people living in the countries where military drones fly. Pakistanis see the drones as a violation of their country's sovereignty. Civilian casualties from missiles either missing their targets or killing innocents nearby. The debate remains whether accidental deaths are much lower with drones than with traditional methods of warfare like airstrikes or ground combat. The U.S. government and military argue that they are. Others question the criteria under which targets are determined and are worried about the ethics and legality of drone strikes.

THE DRONE WAR ON ISIS

More so than in other conflicts, the United States has relied heavily on drones in its assault on ISIS, the militant group in the Middle East that has taken over a large territory in civil war–torn Syria and neighboring Iraq. Operation Inherent Resolve saw the United States and its allies carry out more than 875 drone airstrikes in Iraq and Syria within its first ten months.

GOING GREEN

The Navy is helping the military go green, so to speak. The U.S. military views climate change as a legitimate threat to international security, due to its projected effects on food supplies and the resulting instability and migrations it is expected to cause in the coming decades.

Besides taking troops out of harm's way, drones perform tasks that allow military forces to alleviate the need to move large amounts of personnel into harsh terrain. This has been the case in Afghanistan, for example, where the mountainous landscape has made moving supplies, troops, and weaponry by ground incredibly expensive. The amount of fuel necessary, in addition to other resources, can be astronomical. Fuels and supplies must also be imported into the country to begin with.

With drones taking on more and more of the functions previously handled by actual soldiers, the move to drone technology and away from human resources promises to diminish the environmental impacts of missions, both directly and indirectly.

The commander of the 432nd Wing at Creech Air Force Base, Colonel James Cluff, told Reuters that they "were involved in pretty much every engagement." One such engagement saw the drones ensure that supplies were correctly dropped to tens of thousands of civilians surrounded by ISIS.

Drones and Warfare

Fighters from Isis (the Islamic State Militant group) load what they claim are parts of a drone that crashed near Raqqa, Syria, in the midst of the Syrian civil war in September 2014.

    Predator and Reaper drones have also played vital supporting roles in raids. The drones took part in what the Air Force calls buddy-lasing. This means drones flying ahead of fighter jets and bombers, spotting militant forces with their cameras. They then point their invisible lasers at the targets, and the fighter jets and bombers fire their weapons following the drones' lasers.

SENDING SUPPORT

Drones have also assisted with civilians and military units on the ground surrounded by ISIS militants, among other support missions during war. In December 2011, a K-MAX helicopter drone carried 3,500 pounds (1,588 kg) of food and supplies to US Marines in Afghanistan. The K-MAX takes soldiers' lives out of harm's way, making it an appealing option for the military on resupply missions. It can also reduce the number of truck convoys (often popular targets for bomb attacks) and reduce the amount of cargo that convoys must carry, which can impede their mobility while they are vulnerable.

Unmanned aerial vehicles have the ability to move through areas otherwise impassable due to floods, destroyed roads, lack of infrastructure, or inhospitable terrain. Disaster areas in need of supplies can receive them via drones that are impervious to things dangerous to humans such as fire, smoke, radiation, or toxic chemicals. Additionally, they can also help map out routes to send military relief convoys on or aid points.

CHAPTER 3

THE ETHICS OF DRONE WARFARE

In November 2011, a $6 million secret weapon departed from Shindand Air Base in western Afghanistan. Its mission was to travel hundreds of miles undetected by radar to spy on Iran's nuclear program and discover dangerous activities possibly being engaged in in Afghanistan. The plan was for the drone to circle the target zone silently for hours on end, providing the United States with photos and information no human pilot could retrieve. The drone was sleek, sophisticated, and the end result of fifteen years of development conducted in secrecy. It was cutting edge, and it was the future of warfare. So it came as a surprise to everyone involved when the mission ended in disaster.

For all of its advantages for the military, the drone warfare policies of the U.S. and other nations have drawn harsh

The Ethics of Drone Warfare   31

Pakistani Tribal members from the nation's north Waziristan region—reportedly a hotbed of militant activity—attend the funeral services of several kin slain in a drone attack.

condemnation internationally. The missions do not always go as planned, and there have been unfortunate casualties. Targeted missions have caused many to wonder if drone strikes should be legal, and now many more countries have adopted or will adopt drone technology.

TARGETED MISSIONS

Weaponized drones carry out special missions that have become the subject of controversy. A targeted killing is a

Drones and Warfare

Anwar Al-Awlaki, an American-born US citizen who resettled in Yemen and allegedly was a major figure in Al-Qaeda, was assassinated by a US drone in 2011.

mission carried out by a state or government using lethal force on a human being. There are two types of drone attacks carried out by the United States, mostly under President Obama's administration. One is a personality strike that targets "named, high-value terrorists." The other is a signature strike that targets camps and suspicious compounds in areas known or highly suspected to be hostile.

Missions such as these have been carried out by the military for years but have transformed with the introduction of UAVs. Rather than sending in soldiers to carry out a targeted mission, drones are now sent out in their place. The United States' war on terror has provided the rationale and impetus for the government to increase not only drone usage, but also targeted killings carried out by drones. However, regardless of the success of a mission, there are many who disagree over the usefulness and morality of targeted killings.

SUPPORTERS

Drone-targeted killings have become the weapon of choice in the war on terror beginning with President Bush and continuing on through President Obama. The preference for drone strikes in targeted missions is due to a few reasons. For one, drones do not put the lives of American combat forces on the line. Furthermore, the authorities in nations such as Pakistan and Yemen are very resistant to having American troops in their countries. With better intelligence on its enemies, the United States government can just simply eliminate a target rather than bringing that person in for questioning.

Thus, sending troops into a hazardous area to arrest a target puts too many lives at risk. Also, there could be negative diplomatic repercussions should the United States send soldiers instead of drones. While drones themselves are unpopular in many countries, they are still preferred to having troops on their soil, according to many diplomatic experts who listen to feedback from the nations in question.

According to a poll, drone strikes are a popular option for fighting terrorism among Americans. The research showed that 65 percent of Americans approved of drones strikes that kill suspected foreign terrorists, while 28 percent were opposed. Drone strike supporters assert that there would a public outcry should an American die in a raid in which a drone strike would have done the job.

OPPONENTS

Many oppose targeted killings. For them, the convenience of assassinating rather than capturing a target is simply not enough reason to continue with the practice. One criticism deals with what makes a person or group a target in the first place. What the CIA classifies as a "terrorist signature" may be too broad, critics say. The definition of "suspicious" may not be specific enough. For example, a group of men loading a truck with fertilizer could incorrectly be labeled as bomb makers when they could very well be farmers.

A poll taken in 2012 showed that only 17 percent of Pakistanis favored drone strikes against extremist groups. Publicly, foreign governments disapprove of them, although they prefer them to hosting a foreign military presence. In the United States, members of Congress accuse the Obama administration of preferring drone strikes because he closed down overseas prisons where the targets would be normally sent, and they are avoiding sending more to the controversial detention camp in Guantánamo Bay, Cuba.

Human rights advocates say that the targeted missions are simply extrajudicial killings that run afoul of the law. Furthermore, many believe targeted militants should be given a fair trial and have their identities established rather than being executed without a day in court and dis-

These protesters gathered in November 2013 in Islamabad, Pakistan, to denounce US Drone killings, in particular their frequency in the tribal borderlands it shares with its Western neighbor, Afghanistan.

covering whether they truly pose a threat to the United States or not. Particularly controversial have been the times the United States has used drones to target actual US citizens overseas.

Over the years, the Obama administration has shown that they too may be losing their preference for drone strikes. Drone strikes in Pakistan have actually decreased in number there annually. There were 117 strikes in Pakistan in 2010, 64 in 2011, and 46 in 2012, according to the *New York Times*.

CIVILIAN IMPACT

Drones have an effect beyond those on their intended targets. While drones may take a pilot out of the line of fire, the same is not always true for bystanders. Public opinion on drones varies everywhere. While opponents condemn innocent civilian death, drone attack supporters say that civilian casualties are not as high as opponents claim and save lives in the long run.

SAVING LIVES

Those in support of drone warfare point out that regardless of the weapon, there is no such thing as humane war. Drones are just the best choice from a set of difficult options. Compared to other military weapons, they argue, drones kill fewer civilians. In 2012, the United Nations released a report showing that civilian casualties caused by the United States fell 46 percent from the previous year.

Some deaths attributed to drones were more likely the result of traditional planes and the bombs they carry. The same report showed that using more drones instead of manned airstrikes, while causing 15 more deaths than the year before, saved 124 lives that would have ended from those manned missions. This means moving to drones saved a net total of 109 lives.

Drones shorten the length of time in battle. Sending troops leads to fighting, which can be a prolonged engagement. Long battles lead to more deaths and more risk to innocent civilians. Drones, on the other hand, strike fast. They minimize the time spent fighting and reduce the number of lives at risk.

In past wars, from World War II through the Persian Gulf War, bombs and mortars were responsible for hundreds of thousands of civilian deaths. In Vietnam, aerial bombing killed many thousands of civilians. In Kosovo during the 1990s, planes were ordered to fly no lower than 15,000 feet (4,572 m) to protect civilians. However, in practice this meant that pilots could not accurately identify targets. Drones have fewer limitations. Furthermore, the missiles carried by drones are arguably more accurate than the bombs or missiles delivered by jets. Compared to traditional warfare, their defenders claim, drones are said to save lives in the long term.

CASUALTIES

Some human rights groups claimed that by 2012, more than eight hundred civilians had died in drone strikes. Although the government downplays the risk to civilians, the first major drone success in 2002 also resulted in the death of a US citizen. In 2010, instead of tracking a senior Taliban leader, they instead tracked the cellphone of a human rights advocate and, ironically, a US supporter. The wrong individual was killed

On another mission in October 2010, a drone strike successfully assassinated a leader of the Pakistani Taliban named Qari Hussai. The strike came at the end of a string of

Demonstrators for the antiwar group Code Pink protest the Unmanned Systems 2013 convention in Washington, D.C., which was attended by drone makers and buyers—including the US military and intelligence community.

failed attempts that resulted in a total of 138 deaths, icluding 12 children, according to the *Guardian*.

The official number of civilian casualties is lower than critics claim. However, there have been charges that the Obama administration was dishonest about the number of civilian deaths from drone strikes. The government would consider all military-aged men killed in the drone strike zone as targets, and would classify them as a civilian only if there was evidence of their innocence after the fact. This may account for John Brennan, President Obama's former counterterrorism adviser and current director of the CIA, proclamation that there were no civilian deaths due to drone strike in 2011, according to various news sources, a claim that many critics and observers found dubious.

GLOBAL PROLIFERATION

Since their inception, more countries and governments around the world have adopted drone technologies. This proliferation includes many countries around the world. Ninety countries have drones of some kind, a number of thoseusing them for combat, including Israel, Nigeria, Pakistan, and the United Kingdom. At least a dozen states, such as

Taliban leader Mullah Akhtar Mansour was believed to have been traveling in this now burning car in a remote part of Pakistan.

POTENTIAL FOR MISUSE

Governments are not the only ones using drones in warfare. There have been reports of militants hacking into drones and spying on governments' surveillance videos. Consumer-grade drones that can be purchased by anyone have also been used by terrorists for smuggling drugs, surveillance, and attacks.

In 2013, a German-law enforcement team prevented Islamic militants from carrying out drone attacks. At least six potential terrorist drone attacks have been stopped since 2011 in the United States, Germany, Spain, and Egypt. Security concerns have been raised by hobbyist drone pilots—operating with no ill intent—who showed holes in security by flying their UAVs into vulnerable locations, such as a stadium filled with people or to the White House.

Hamas, a group considered an enemy force by the United States and Israel, possesses the Iranian-made Ababil-1 drone. The Ababil-1, which can travel 125 miles per hour (201 kilometers per hour) with a ceiling of 5,000 meters (16,404 feet), can launch missiles as well as nose-dive into targets. However, experts believe that these drones do not represent a power shift in the area, and security measures can be taken against them.

In the United States, many are concerned with local police forces acquiring their own drones. Only fourteen states have laws limiting drone use to protect privacy. Police departments are otherwise free to use drones in ways that privacy advocates find troubling. While some police departments, such as Mesa County Sheriff's Office in Colorado, have strict rules in place for drones, other departments are not as disciplined. Most police departments use drones only for search-and-rescue missions, crime scene photography, or assisting bomb technicians. But with drone regulation still in its infancy, the possibilities for drone misuse exists.

Iran, Iraq, China, and Saudi Arabia, are believed to possess armed drones, while other countries, such as India, are in the process of acquiring or developing them.

The United States has sold unarmed drones to Spain and the Netherlands and agreed to weaponize drones for Italy's military. Other European countries are developing them, too. Terrorists and militants in various countries are also thought to be using drones for their own purposes.

There are many who believe that drones make wars less expensive and therefore more likely to occur. Drone proliferation, in their eyes, is unavoidable and represents significant risks to security.

Others disagree and see drones as nothing more than an updated military technology, with drone proliferation a low priority. The US government could either try to regulate drone exports to minimize their spread or be more willing to arm its allies. Not all countries are dependent on the United States, however. China is a willing drone exporter, with Nigeria, Pakistan, and Iraq as buyers. The technology used in most drones does not match the capabilities of US-owned drones, so many countries are years away from operating drones on the same level as the US military.

INTERNATIONAL LAW

The legality of drone strikes has been legitimized by redefining what a battlefield is. The US Constitution requires the government to allow a suspect who has been arrested a fair trial. However, due process, or the requirement of a fair trial, can be skipped when it comes to warfare and fighting on a battlefield. With the battlefield redefined to virtually most of the world, the United States is able to find and kill their targets anywhere they are found.

President George W. Bush and his administration set the precedent that criminal and international humanitarian laws did not apply to targets in the war on terror—which remains a controversial concept, as mentioned. The government points to the Authorization for the Use of Military Force passed by Congress soon after the terrorist attacks on September 11. It states that the president may "use all necessary and appropriate force against those nations, organizations, or persons . . . in order to prevent any future actor of international terrorism."

The British government's Ministry of Defence released a report that looks ahead to the year 2035. The consensus was that international law will not be able to keep up with the rate of change in technology to protect ethical and moral considerations. Stealth drone technology could be outlawed based on a violation of the Geneva Conventions, which define standards for war crimes for all nations. But that is just the beginning of new technologies.

CHAPTER 4

Drone Warfare of the Future

Drones have come a long way since their inception. To someone from an earlier era, laser-guided missiles launched by drones with pilots half a world away may have sounded like pure magic. What kind of future is in store for drones and drone warfare? Will technology continue to improve in leaps and bounds? And will our legal frameworks be able to effectively adapt?

DRONE CARRIERS AND FUTURE CRAFT

Cameras and sensors on drones far exceed the capabilities of their predecessors of even just a few years ago.

A French soldier stands guard near a Reaper drone in Niamey, Niger, part of the resources France and other nations have dedicated to fighting Islamic militants in Africa.

Video surveillance systems such as the Autonomous Real-Time Ground Ubiquitous Surveillance-Imaging System, or Argus-IS, provides real-time video streams that can track people and vehicles from 20,000 feet (6,096 m) in the air across an area of approximately 65 square miles (105 sq km). This system is getting upgraded and will offer night vision and infrared imaging sensors.

    The next step in drone warfare includes the UXV Combatant, a concept warship developed by British defense company BAE Systems set for launch after 2020. The UXV will be about 500 feet (152 m) long and run on diesel-powered electric turbines. It is planned as a command and control center for a variety of unmanned vehicles. The warship will be run with a fraction of the manpower necessary for traditional battleships. While battleships carry hundreds of sailors, the UXV will be designed with with a sixty-person crew in mind, along with whatever personnel are necessary to control and maintain the drones on board.

    In addition to next-generation weapons, the UXV will be the launch point for at minimum twenty-four military drones stored on the ship. Electromagnetic catapults are designed to launch drones from adjustable ramps located on the deck. Representing another level of organization, radar, sensors, and radio-frequency identification chips are used to coordinate multiple drones at once. The UXV will be able to carry out a wide range of missions, including submarine hunting, mine sweeping, and providing supplies to troops. As a stealth ship, the UXV could quietly patrol the seas and carry out missions at a moment's notice.

MICRODRONES AND SWARMS

Much of the innovation in drones happening currently and looking to the future can be summed up in two words: think smaller. The PD-100 Black Hornet is one of the latest drone designs that look to carry out missions differently from their larger counterparts. This microdrone is the size of a hummingbird, fits in the palm of a soldier's hand, and weighs only .63 ounces (18 grams). The Black Hornet can move ahead of troops and help detect any threats lurking around a corner or spy on an enemy sniper hidden away.

    The microdrone, called a personal reconnaissance system, is also relatively easy to use. It has a flying range of 1 mile (1.6 km), can hover for close to half an hour, and can fly to heights as high as 10,000 feet (3,048 m). On the Black Hornet are three cameras facing different directions that can take video or still photos. It has already been deployed with Marine Corps special operations teams. A Marine veteran said that the microdrone could save lives in combat since the drone can be sent in to carry out dangerous jobs such as inspecting vehicles for explosives. The Black Hornet is small enough that a Marine can carry two and use one while the other charges its battery.

    Since microdrones are relatively cheap, the next advance in the technology is to use entire swarms of them for various missions. The US Navy is set to test this idea in the summer of 2016 as part of its Low-Cost UAV Swarming Technology, or LOCUST, program. They will set out thirty drones over the ocean to test their capabilities flying together. In a big technological breakthrough, each drone in a swarm is aware of one another's movements.

This Covert Autonomous Disposable aircraft, or Cicada, represents the likely future of drone technology—smaller and more stealthy, but as powerful in its surveillance abilities as older, bigger drones.

 The advantage of a swarm is that if one drone is brought down by hostile fire or a technology issue, the rest of the swarm can adjust, compensate, and continue the mission. While a large drone can sustain only so much damage before failing, a swarm can withstand multiple hits or individual losses, making it difficult to stop, especially by defense systems designed to take down a single aircraft. The swarm can simply overwhelm an enemy. The equipment on each swarm drone, from cameras to radar jammers, is inexpensive to reduce overall cost and heighten the drone's expendability. However, some are hesitant to incorporate swarm drone technology. While one autonomous drone gives many critics reason for concern, an entire swarm acting largely on their own makes policymakers cautious.

INTELLIGENT MACHINES

By February 2015, the military had more than seven thousand unmanned vehicles in the air. The number of drones on the ground was even more: twelve thousand, including iRobot's PackBots, which searched for bombs along the roads in Afghanistan.

Today, unmanned vehicles still require a human in control somewhere either near or far away. In the future, unmanned vehicles could operate completely free of human control. The most technologically advanced drones today have some autonomous features, such as the ability to take

The crew here were the first to catapult-launch a drone from the deck of an aircraft carrier. This X-47B Unmanned Combat Air System (UCAS) took flight on May 13, 2013, on the Atlantic Ocean.

off or land on their own, and even to fly to predetermined locations. Eventually, entire missions will be carried out by technology alone.

The Northrop X-47 test aircraft was able to take off and land on an aircraft carrier, a task typically difficult to execute for human pilots. The British military is testing a jet-powered drone called Taranis that is not only a stealth model that can fly itself, but it can find and select what targets to attack.

LOW-COST PROTECTION

One trait of technology is that it gets cheaper with each passing generation. A computer today will cost less next year. The same can be said about drone technology. For the military, drones are a cheaper option than using traditional force, which risks their pilots' lives. Drones also provide governments with a tool to protect their own countries and increase security at a relatively low cost, at least beyond initial investments in the technologies. Experts believe this reduces the potential for upsetting the status quo, or the current state of affairs, of a country. Domestic drones may prove to be more useful than those used in combat since the technology in use today is vulnerable to air defense systems. However, experts in drone technology also warn that the world's governments must be careful not to escalate a small skirmish into a war by launching drones because it is cheap to do so.

The low cost of drone protection may turn into something much more sinister. Unmanned vehicles could be an attractive option for leaders of authoritarian countries, who can turn the drones on their own citizens. Local populations could be subjected to domestic repression.

In 2014, China used drones for surveillance purposes against local protesters in Xinjiang. If a dictator had armed drones, then he would be able to conduct more than just surveillance missions. Former Iraqi leader Saddam Hussein was not able to fight both domestic threats and foreign threats, such as the US military. With cheap drones, he may have been able to defend his authority better. Interestingly, dictators require a strong military to maintain control, yet it is this same military that has the power to overthrow the dictator. Drones could make a dictator "*coup d'état*-proof" (that is, protected from being overthrown) by reducing the size of the military but still maintaining power through a fleet of robotic soldiers.

RESHAPING WARFARE

Unmanned vehicle technology is moving forward. Drones are filling up the sky, the ground, and the seas with models that fly through caves and unmanned submarines that navigate the deep waters of the oceans. Where one large drone used to suffice, a swarm of microdrones can be sent in to do the job at a more efficient and cost-effective pace. The effect of the advancing technology has global implications.

In the United States, drones have become an acceptable form of warfare because they take human personnel out of harm's way. Antiterrorism efforts are increasingly

This close-up shows a powerful Hellfire missile hanging from the bottom of the USAF's MQ-1B Predator Drone, parked at a secret air base somewhere in the Persian Gulf region in early 2016.

fought by drones that seek out and target militants. These drones have proven to be effective, but at the same time not entirely reliable as many unintended targets have fallen victim to various faults in the technology or human error. Despite the myriad critics calling for an end to the unmanned vehicle practice, the military continues to utilize drones and expand their utility. And the United States is not the only country flying drones into combat zones.

As models get cheaper, drone proliferation will become more prevalent, according to many experts. This has already been observed as many countries around the world either currently have drones or are in the process of building their own. While the US military may have an advantage by having the most advanced models, other countries are racing to catch up and perhaps even surpass American technology.

China is quickly becoming a major factor in the drone economy, and effectively, a major factor in modern warfare. They are focusing many resources on building stealth drones that can fly undetected, forcing defense systems to adapt to the new threat. Meanwhile, legal battles are being fought to determine the legality of drones in general. Humanitarians decry the deaths of innocent civilians, while others point to the lives saved in drone attacks, as opposed to traditional troops battling on the ground.

So what will warfare in the future look like? Could it be a winner-take-all battle of robots? Will the future be as awe-inspiring to us as self-flying drones would be to a hot-air balloon pilot from the Civil War? The technology is advancing every day, and today's expensive breakthroughs will eventually become yesterday's cheap alternatives. This is already seen in today's drone market, as more competitors drive the prices down while more features are added.

Hollywood movies have made a fortune predicting a future in which machines have taken over and enslaved the human race, as the scientists who created the war machines regret their inventions. Or will wars be human-casualty free, where metal, plastic, and circuit boards are the only things at risk? Experts believe that these predictions are still too far off based on where the technology and global

An attendee examines some Chinese-made drones at a drone exhibition in Beijing. China is already a pioneer in commercial drones, and is expected to become competitive in building military drones soon.

marketplace is today. The focus, they argue, should be on the impact of drones today. However, what happens today could have a long-lasting effect on what we face in the future and where the human race will go.

GLOSSARY

ACQUISITION The act of assuming possession of something by one's own efforts.
ARTILLERY Large firearms such as cannon or rockets.
CONSENSUS The judgment arrived at by most of those concerned.
COUNTERTERRORISM Military activities that are engaged in to prevent and combat terrorism.
COUP D'ÉTAT A sudden attempt by a group to take over a country's government, often by violent means.
DIPLOMATIC Relating to, or concerned with the work of keeping up relations between the governments of different countries.
ESPIONAGE The practice of spying or the use of spies.
EXTRAJUDICIAL Happening out of the court system or not legally authorized.
GLOBAL POSITIONING SYSTEM A satellite-based navigation system made up of a network of satellites placed into orbit by the US Department of Defense.
HOBBYIST Someone who has an interest or activity to which he or she devotes leisure time.
IMPERVIOUS Not disturbed or upset or unable to be affected by.
MILITANT A person aggressively active, especially in a cause.
MORALITY The quality of, or relating to, the judgment of right and wrong in human behavior.
MUNITIONS Military ammunition, weapons, or equipment.
PROLIFERATE To grow or increase rapidly or the rapid increase in numbers.
RECONNAISSANCE A survey (as of enemy territory) to gain information.
REPRESSION The act of putting down by force.
SOVEREIGNTY Freedom from outside control or a self-governing state.
STATUS QUO The existing state of affairs.
UBIQUITOUS Existing or being everywhere at the same time.

FOR MORE INFORMATION

Academy of Model Aeronautics (AMA)
5161 E. Memorial Dr.
Muncie, IN 47302
(800) 435-9262
Website: http://www.modelaircraft.org
Established in 1936, the AMA is the world's largest model aviation association whose mission is to promote the development of model aviation as a recognized sport and recreation activity. The AMA can show you where to find instructors, how to become involved, and answer questions about model aviation.

Association for Unmanned Vehicle Systems International (AUVSI)
2700 South Quincy Street
Suite 400
Arlington VA 22206
(703) 845-9671
Website: http://www.auvsi.org
The Association for Unmanned Vehicle Systems International (AUVSI) is a nonprofit organization that promotes unmanned systems and robotics technologies. Members are provided networking and development opportunities in the unmanned systems community.

Canadian Centre for Unmanned Vehicle Systems
#4, 49 Viscount Avenue SW, Medicine Hat
Alberta, Canada T1A 5G4
(403) 488-7208
Website: http://www.ccuvs.com
Established in 2007, the Canadian Centre for Unmanned Vehicle Systems is a non-profit organization that promotes growth for the unmanned systems industry and expands knowledge and awareness among the general public in Canada.

Code Pink
1241 Evarts Street, NE
Washington DC 20018
(202) 248-2093

Website: http://www.codepink.org/ground_the_drones
Code Pink is a worldwide network promoting peace and social justice. They are involved with numerous issues, including the use of unmanned aerial vehicles at home and overseas.

Federal Aviation Administration
800 Independence Avenue, SW
Room 908
Washington, DC 20591
(202) 267-3883
Website: http://www.faa.gov
This governmental division's mission is to provide the safest, most efficient aerospace system in the world.

Smithsonian National Air and Space Museum
Independence Ave at 6th Street, SW
Washington, DC 20560
(202) 633-2214
Website: http://airandspace.si.edu
The Smithsonian National Air and Space Museum is the world's largest collection of aviation and space artifacts. The *Military Unmanned Aerial Vehicles* exhibition has several drones through history on display.

United States Department of Defense
1400 Defense Pentagon
Washington, DC 20301-1400
(703) 571-3343
Website: http://www.defense.gov
The US Department of Defense is in charge of the military and all of its branches, helping to protect the country with whatever means available, including unmanned aerial vehicles.

Unmanned Systems Canada
PO Box 81055
Ottawa Ontario K1P 1B1

For More Information

(613) 526-5487
Website: https://unmannedsystems.ca
Unmanned Systems Canada is a nonprofit association that represents the best interests of unmanned vehicle communities among the industry, academia, government, and military and promotes education and appreciation.

WEBSITES

Because of the changing number of internet links, Rosen Publishing has developed an online list of websites related to the subject of this book. This site is updated regularly. Please use this link to access this list:

http://www.rosenlinks.com/IWD/war

FOR FURTHER READING

Alexander, T. M. *Hacked*. Gosford, Australia: NSW Scholastic Australia, 2014.

Calhoun, Laurie. *We Kill Because We Can: From Soldiering to Assassination in the Drone Age*. London, United Kingdom: Zed Books, 2015.

Cheng, Eric. *Aerial Photography and Videography Using Drones*. Berkley, CA: Peachpit Press, 2015.

Collard, Sneed B. *Technology Forces: Drones and War Machines*. Vero Beach, FL: Rourke Educational Media, 2013.

Dougherty, Martin. *Drones: An Illustrated Guide to the Unmanned Aircraft That Are Filling Our Skies*. London, England: Amber Books, 2015.

Faust, Daniel R. *Military Drones*. New York, NY: PowerKids Press, 2016.

Finn, Denny Von. *Predator Drones*. Minneapolis, MN: Bellwether Media, 2013.

Gerdes, Louise I. *Drones*. Detroit, MI: Greenhaven Press, 2014.

Greenhaven Press. *Drones: Current Controversies*. Farmington Hills, MI: Greenhaven Press, 2016.

Hamilton, John, and Sue Hamilton. *UAVs: Unmanned Aerial Vehicles*. Minneapolis, MN: ABDO Publishing, 2012.

Hustad, Douglas. *Discover Drones*. Minneapolis, MN: Lerner Publications, 2016.

Kilby, Terry, and Belinda Kilby. *Getting Started with Drones: Build and Customize Your Own Quadcopter*. San Francisco, CA: Maker Media, 2015.

Mahaney, Ian F. *Extreme Unmanned Vehicles*. New York, NY: PowerKids Press, 2016.

Marsico, Katie. *Drones*. New York, NY: Children's Press, 2016.

Mooney, Carla. *Pilotless Planes*. Chicago, IL: Norwood House Press, 2011.

Nagelhout, Ryan. *Drones*. New York, NY: Gareth Stevens Publishing, 2013.

Nardo, Don. *Drones*. Greensboro, NC: Morgan Reynolds Publishing, 2014.

Norris, Donald. *Build Your Own Quadcopter*. New York, NY: McGraw-Hill Education, 2014.

Otfinoski, Steven. *Drones: Science, Technology, and Engineering*. New York, NY: Children's Press, 2016.

Rauf, Don. *Getting the Most out of Makerspaces to Build Unmanned Aerial Vehicles*. New York, NY: Rosen Publishing Group, 2015.

Ray, Michael. *History of Air Forces Around the World*. New York, NY: Britannica Educational Publishing, 2014.

Rauf, Don. *Getting the Most Out of Makerspaces to Build Unmanned Aerial Vehicles*. New York, NY: Rosen Publishing Group, 2014.

Ripley, Tim. *Drone Operators: Military Jobs*. New York, NY: Cavendish Square Publishing, 2015.

For Further Reading

Rose, Simon. *Drones*. New York, NY: AV2 by Weigl, 2015.

Rupprecht, Jonathan. *Drones: Their Many Civilian Uses and the U.S. Laws Surrounding Them*. Seattle, WA: Amazon Digital, 2014.

Samuels, Charlie. *Machines and Weaponry of the Gulf War*. New York, NY: Gareth Stevens Publishing, 2013.

Sheen, Barbara. *Cutting Edge Military Technology*. San Diego, CA: ReferencePoint Press, 2016.

Simons, Lisa M. B. *U.S. Air Force by the Numbers*. North Mankato, MN: Capstone Press, 2014.

Turner, Tracey. *20 Weapons of War*. Mankato, MN: Smart Apple Media, 2015.

Wesselhoeft, Conrad. *Dirt Bikes, Drones, and Other Ways to Fly*. Boston, MA: HMH Books for Young Readers, 2015.

BIBLIOGRAPHY

Axe, David. "Drones Take Over America's War on ISIS." Daily Beast, June 7, 2015. http://www.thedailybeast.com/articles/2015/06/17/the-war-on-isis-is-a-drone-war.html.

Brandon, John. "What's It like to Pilot a Drone? A Lot like 'Call of Duty'." Fox News, November 26, 2012. http://www.foxnews.com/tech/2012/11/26/whats-it-like-to-pilot-drone-lot-like-call-duty.html.

Byman, Daniel L. "Why Drones Work: The Case for Washington's Weapon of Choice." The Brookings Institution, July/August, 2013. http://www.brookings.edu/research/articles/2013/06/17-drones-obama-weapon-choice-us-countert-errorism-byman.

Callam, Andrew. "Drone Wars: Armed Unmanned Aerial Vehicles." *International Affairs Review*, Winter 2010. http://www.iar-gwu.org/node/144.

Carrington, Damian. "US Military Combines Green and Mean to Fly Drones on Biofuels." *Guardian*, March 15, 2012. http://www.theguardian.com/environment/damian-carrington-blog/2012/mar/15/biofuels-us-navy-drone.

Cohn, Marjorie. *Drones and Targeted Killing: Legal, Moral, and Geopolitical Issues*. Northampton, MA: Interlink Publishing, 2014.

Cuadra, Alberto, and Craig Whitlock. "How Drones Are Controlled." *Washington Post*, June 20, 2014. http://www.washingtonpost.com/wp-srv/special/national/drone-crashes/how-drones-work.

Dorn, A. Walter. *Air Power in UN Operations: Wings for Peace*. Farnham, United Kingdom: Ashgate Publishing, 2014.

Hastings, Michael. "The Rise of the Killer Drones: How America Goes to War in Secret." *Rolling Stone*, April 16, 2012. http://www.rollingstone.com/politics/news/the-rise-of-the-killer-drones-how-america-goes-to-war-in-secret-20120416.

Hennigan, W. J. "Drone Helicopter Brings Supplies to U.S. Troops in Afghanistan." *Los Angeles Times*, January 10, 2012. http://articles.latimes.com/2012/jan/10/business/la-fi-0110-drone-helicopter-20120110.

Leach, Adam. "Robots to the Rescue." *MINE*, May 21, 2014. http://www.nridigital.com/mine-digital-magazine.html?wv=s/MINE%20Magazine/209227db-e351-4b31-8756-039f163dc850/MINE%201405/robots.html.

Nicas, Jack. "Criminals, Terrorists Find Uses for Drones, Raising Concerns." *Wall Street Journal*, January 28, 2015. http://www.wsj.com/articles/criminals-terrorists-find-uses-for-drones-raising-concerns-1422494268.

BIBLIOGRAPHY

Norton-Taylor, Richard. "New Weapons Systems May Breach International Law." *Guardian*, March 15, 2016. http://www.theguardian.com/news/defence-and-security-blog/2016/mar/15/new-weapons-systems-may-breach-international-law.

Pumphrey, Clint. "How Drone Strikes Work." HowStuffWorks, Accessed March 21, 2016. http://science.howstuffworks.com/drone-strikes.htm.

Shane, Scott. "Targeted Killing Comes to Define War on Terror." *New York Times*, April 07, 2013. http://www.nytimes.com/2013/04/08/world/targeted-killing-comes-to-define-war-on-terror.html.

Sifton, John. "A Brief History of Drones." *Nation*, February 7, 2012. http://www.thenation.com/article/brief-history-drones.

Singer, Peter Warren. "The Future of War Will Be Robotic." CNN, February 23, 2015. http://www.cnn.com/2015/02/23/opinion/singer-future-of-war-robotic.

Woods, Chris. "The Story of America's Very First Drone Strike." *Atlantic*, May 30, 2015. http://www.theatlantic.com/international/archive/2015/05/america-first-drone-strike-afghanistan/394463.

INDEX

A
Afghanistan, 4, 10, 14, 20, 25—27, 29—30, 35, 48
Al Qaeda, 10, 20, 24, 32
al-Awlaki, Anwar, 32
Argus-IS, 45
army, 7, 14, 16, 21, 23

B
B-24s, 8
Balkans, Yugoslavia, 19
Bin Laden, Osama, 10
Brennan, John 38
Bush, George W., 24—25, 33, 42

C
China, 2, 7, 41, 50, 52—53
Civil War, American, 8, 18, 52
Code Pink, 38
Cold War, 8
Creech Air Force Base, 12, 27

D
Drones, types
    Ababil-1, 40
    ACTUV, 13
    Black Hornet, 46
    CICADA, 47
    Fire Scout, 16—17
    Global Hawk, 9, 14—15
    Gray Eagle, 16
    Hunter, 16
    LOCUST, 46
    Microdrone, 46, 50
    PackBots, 48
    Phantom, 17
    Pioneer, 10, 16
    Predator, 10, 14—16, 19—21, 25, 28, 51
    Reaper, 5, 15—16, 28, 44
    UAV, 4, 6, 9, 32, 40, 45
    UCAS, 48
    UXV Combatant, 45
    Swarm 46—47, 50
drone strikes, 24, 26, 31, 33—35, 37—38, 42

E
ethics, 26, 30—31, 33, 35, 37, 39, 41

F
First Gulf War, 10, 19, 21, 37, 52

G
Geneva Conventions, 42
Germany, 8, 40
GPS, 11
Guantánamo, 34
Guardian, 38

H
Hellfire II missiles, 16, 51
hot-air balloons, 8, 18
Hussai, Qari, 37
Hussein, Saddam, 50

I
Iran, 30, 40-41
Iraq, 10, 23, 26, 41, 50
ISIS, 26—29

INDEX

Israel, 10, 39—40

K
Kennedy, John F., 19
Kosovo, 37

M
Mansour, Mullah Akhtar, 39

N
Niger, 39, 44
Nigeria, 39
9/11, 10

O
Obama, Barack, 24, 32—35, 38
Operation Inherent Resolve, 26
Omar, Mullah Mohammed, 20—21

P
Pakistan, 24—26, 31, 33—35, 37, 39, 41

Pentagon, 10

R
remote-controlled, 8, 21—22

S
al-Sadar, Muqtada, 23
Salon, 23
Saudi Arabia, 41
Shindand Air Base (AF), 30
Soviet Union, 8
Spain, 40—41
Spanish-American War, 8
Syria, 26, 28

T
Taliban, 20, 24, 37, 39

U
United Kingdom, 39
United States, 4—6, 8, 10, 12, 18—20, 26, 30, 32—36, 39, 40—42, 50—51
    Air Force 4, 12, 14—17, 19—21, 26—28, 42, 49
    Constitution 42
    Department of Defense 10, 15, 19

V
Vietnam War, 8, 10, 37

W
War on Terror, 6, 10, 23, 32—33, 42
World War I, 7, 8, 18
World War II, 8, 18, 37

Y
Yemen, 24—25, 32—33

ABOUT THE AUTHOR

Jeff Mapua has written several books on technology and vessels with titles including *Making the Most of Crowdfunding*, *Extreme Motorsports*, and *A Career in Customer Service and Tech Support*. From radio-controlled cars in his childhood, to consumer drones and helicopters in his adult life, Mapua has always been fascinated with the future of robotic technology. Mapua lives in Dallas, Texas, with his wife, Ruby.

PHOTO CREDITS

Cover, p. 1 Colin Anderson/Photographer's Choice/Getty Images; p. 5 Isaac Brekken/Getty Images; p. 9 USAF/Hulton Archive/Getty Images; p. 12 The Washington Post/Getty Images; p. 14 © Uber Bilder/Alamy Stock Photo; p. 16 Jung Yeon-Je/AFP/Getty Images; p. 20 Getty Images; p. 21, 51 John Moore/Getty Images; p. 23 Ahmad Al-Rubaye/AFP/Getty Images; p. 25 Mohammed Huwais/AFP/Getty Images; p. 28 AFP/Getty Images; p. 31 Thir Khan/AFP/Getty Images; p. 32 The Washington Post/Getty Images; p. 35 Aamir Qureshi/AFP/Getty Images; p. 38 Saul Loeb/AFP/Getty Images; p. 39 AFP/Getty Images; p. 44 Pascal Guyot/AFP/Getty Images; p. 47 Laurent Barthelemy/AFP/Getty Images; p. 48 MC2 Timothy Walter/U.S. Navy/Getty Images; p. 53 Greg Baker/AFP/Getty Images; cover and interior page backgrounds (geometric patterns) Sumkinn/Shutterstock.com; interior pages backgrounds (sky) Serg64/Shutterstock.com, (wave pattern) Kwirry/Shutterstock.com.

Designer: Brian Garvey; Editor/Photo Researcher: Philip Wolny